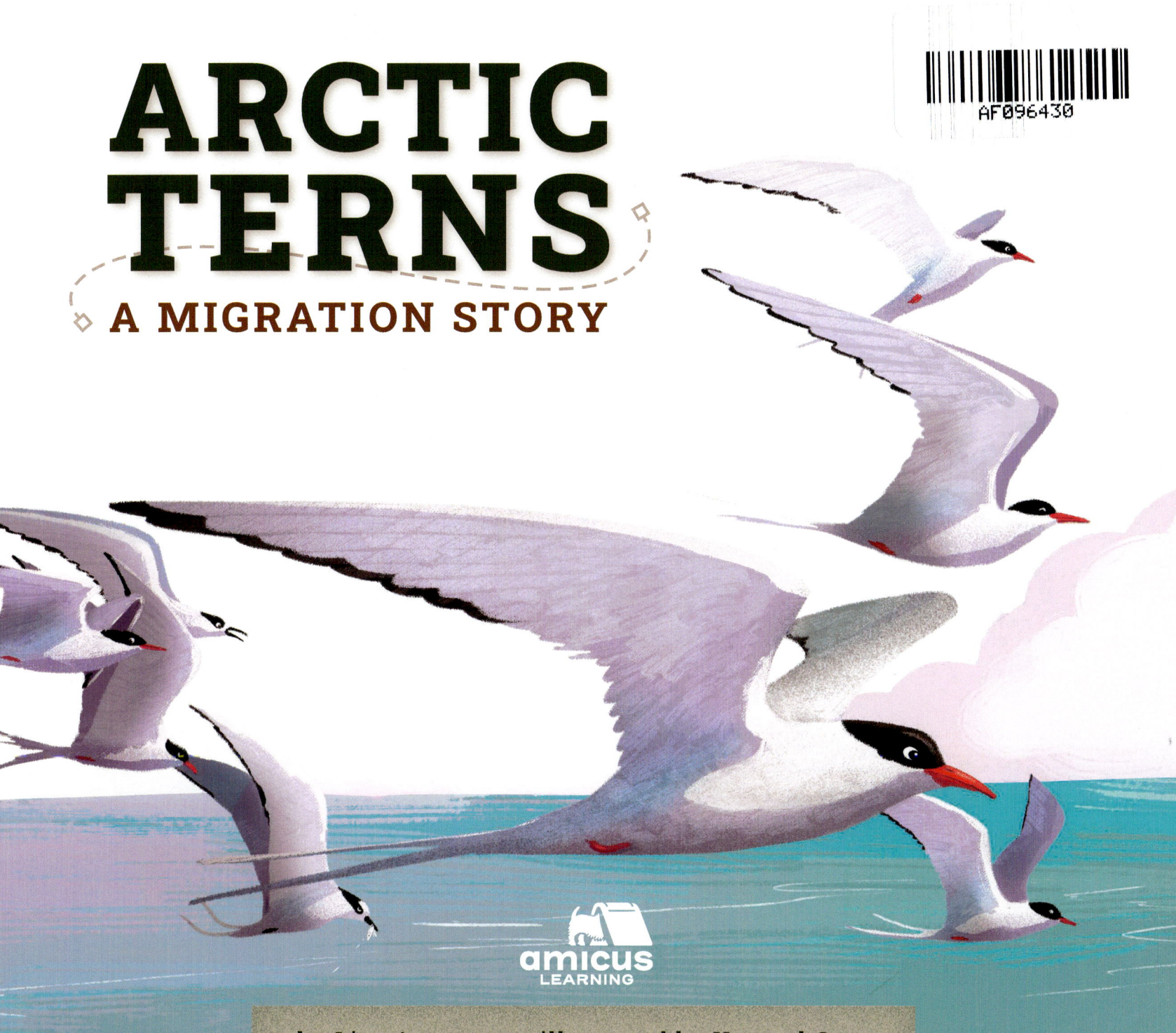

# ARCTIC TERNS
## A MIGRATION STORY

by Lisa Amstutz • illustrated by Howard Gray

amicus LEARNING

AMICUS ILLUSTRATED is published by Amicus Learning, an imprint of Amcius P.O. Box 227, Mankato, MN 56002
www.amicuspublishing.us

COPYRIGHT ©2026 Amicus. International copyright reserved in all countries. No part of this book may be reproduced in any form without written permission from the publisher.

Editor: Alissa Thielges
Series Designer: Kim Pfeffer
Book Designer: Emily Dietz

Library of Congress Cataloging-in-Publication Data
Names: Amstutz, Lisa J., author. | Gray, Howard, illustrator.
Title: Arctic terns : a migration story / by Lisa Amstutz ; illustrated by Howard Gray.
Description: Mankato, MN : Amicus Illustrated, [2026] | Series: Incredible migrations | Includes bibliographical references. | Audience: Ages 5–10 | Audience: Grades 2–3 | Summary: "Follow the incredible journey of an arctic tern's worldwide migration between the Arctic and Antarctic in this narrative nonfiction picture book that will delight animal and nature lovers and support life science education. Includes a migration map, tips to protect animals and habitats, a glossary, and further resources"—Provided by publisher.
Identifiers: LCCN 2024043354 (print) | LCCN 2024043355 (ebook) | ISBN 9781645492801 (library binding) | ISBN 9781681528045 (paperback) | ISBN 9781645493686 (ebook)
Subjects: LCSH: Arctic tern—Juvenile literature. | Arctic tern—Migration—Juvenile literature.
Classification: LCC QL696.C46 A57 2026 (print) | LCC QL696.C46 (ebook) | DDC 598.3/38—dc23/eng/20250107
LC record available at https://lccn.loc.gov/2024043354
LC ebook record available at https://lccn.loc.gov/2024043355

## About the Author
Lisa Amstutz is the author of more than 150 children's books. A former outdoor educator, she holds degrees in biology and environmental science. Lisa enjoys learning fun facts about science and sharing them with kids. She lives on a small farm with her family.

## About the Illustrator
Howard Gray has illustrated a selection of fiction and non-fiction children's books. He has always considered himself an artist, but with a PhD in dolphin genetics, he has a background in zoology. He is now pursuing his dream career in children's illustration from the picturesque city of Durham, UK. Find out more at www.howardgrayillustrations.com.

*Cheep, cheep!* Two fuzzy chicks beg for food. Their mother flies off to find them a fish. It is spring—nesting season for Arctic terns. These birds live in northern Canada and Alaska, not far from the North Pole.

The chicks grow fast. After a few weeks, they learn to fly and find food.

Soon dark, cold days lie ahead. The family must migrate before winter comes. They will travel to Antarctica, near the South Pole. There they will find food and warmer seas.

*Flap, flap!* The mother tern stretches her wings. Her noisy colony grows silent. Then all at once they lift off. Their long journey has begun. They will follow a zig-zag path. They will fly up to 60,000 miles (96,000 kilometers).

Ocean breezes carry the tern. She soars day and night, napping as she flies. When she is hungry, she snatches a fish or bug.

Back home, foxes, cats, and larger seabirds prey on terns. Here, there are few predators. But there is still danger ahead.

*Boom!* Thunder crashes. High winds batter the tern. Down, down she tumbles. She lands in the rough sea. She bobs in the waves, waiting for the storm to pass.

Climate change is making these storms bigger each year. The warmer oceans form bigger storms. They can blow the birds off course.

At last, the storm calms. The tern is weak and hungry. She lifts off to scan for food.

*Swoop!* **The tern dives toward the water. She scoops up a small fish. Uh-oh! A gull steals her catch. The tern dives again and again until her belly is full.**

The flock flies on and on. After many miles (kilometers), they grow tired. They stop to feed and rest their wings. It is the first of several long rest stops to come. Some are along the coast. Others are at sea.

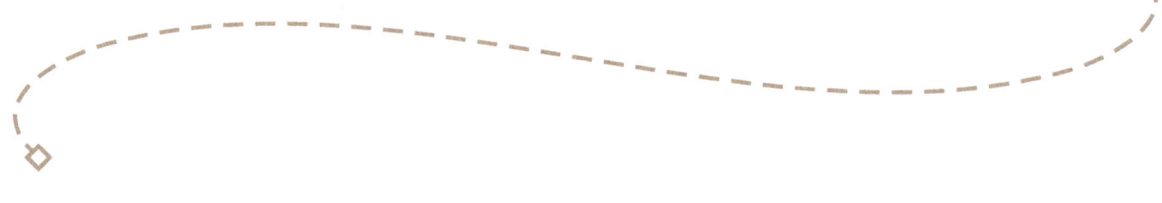

At long last, the tern reaches Antarctica. It has been four months since she left home. It is summer here. The days are long. The tern settles on the pack ice to feed.

The ocean teems with fish. But there are fewer each year. Ice at the poles is melting. The warming seas drive away fish.

While the tern rests, she molts. Her bill and legs turn black. Her feathers fall out and grow back. They will be fresh and strong for the trip home.

Months pass. Summer in the southern hemisphere ends. Dark, cold days lie ahead. It is time to move on.

The tern is rested and ready. Her new feathers shine. *Flap, flap!* She stretches her wings. The air grows silent. Then the colony lifts off . . .

to begin their long journey home.

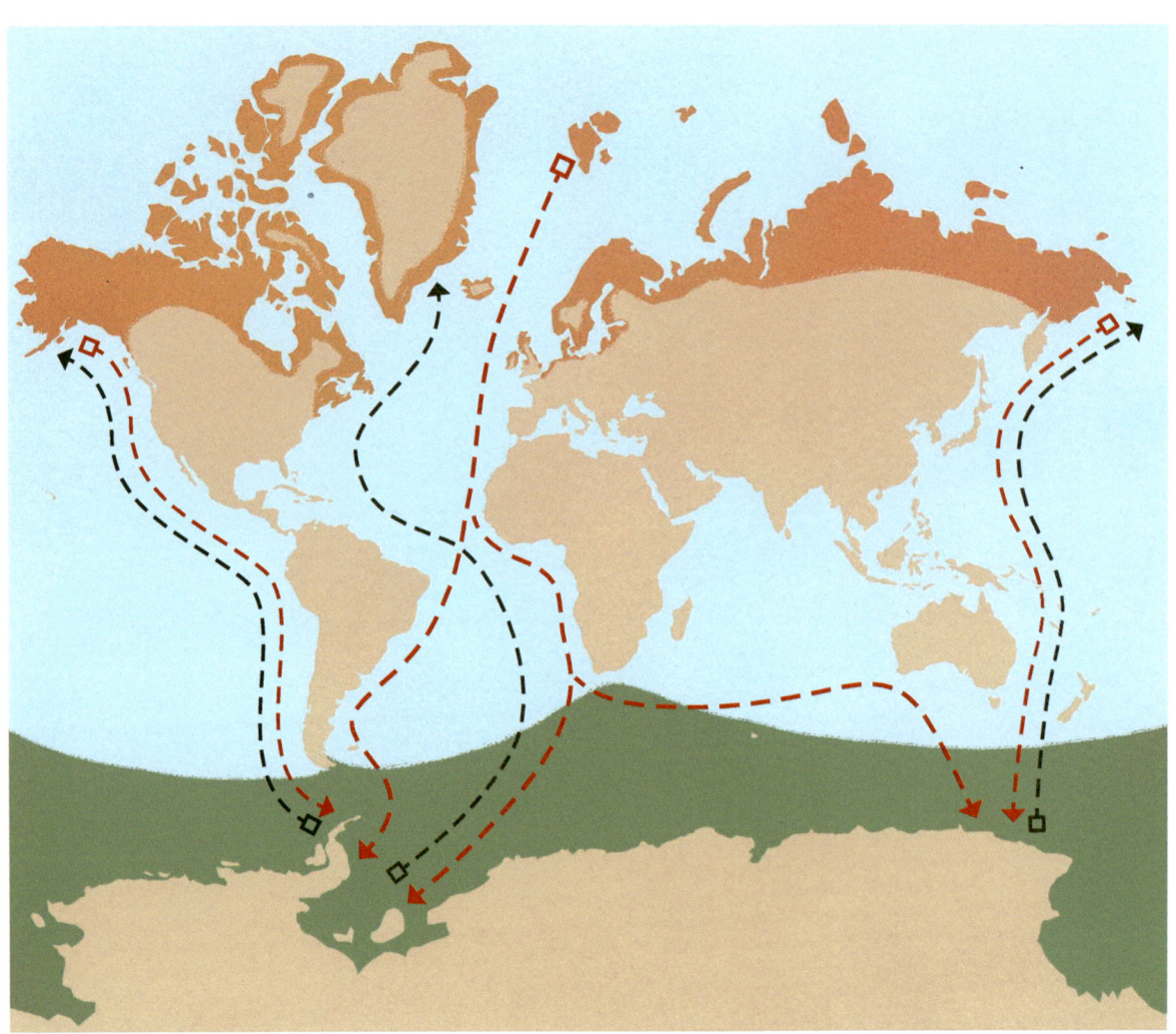

## HOW CAN I HELP ARCTIC TERNS?

- Use fewer plastic products. Plastic can end up in the ocean and harm sea life.

- Stay away from nesting birds. Watch them from a safe distance.

- Buy fish that are caught without harming the environment. The Monterey Bay Aquarium's Seafood Watch provides a helpful guide: https://www.seafoodwatch.org/recommendations/download-consumer-guides

- Bike and walk to places when you can. This uses less fuel and can help slow climate change.

- Spread the word. Let people know how they can help, too.

## Glossary

**climate change** A long-term change in global or regional weather patterns.
**colony** A group of animals that live together.
**hemisphere** One half of the Earth.
**migrate** To move to a different place at a certain time of year.
**molt** In birds, to lose feathers so that new ones can grow.
**pack ice** Pieces of floating ice frozen together to make a very large sheet.
**predator** An animal that hunts other animals for food.

## Read More

Banks, Rosie. *Why Do Animals Migrate?* New York: Gareth Stevens Publishing, 2024.

Loewen, Nancy. *My Arctic Tern Migration Journey.* North Mankato, Minn.: Picture Window Books, 2025.

Musso, Timothy. *Chasing the Sun.* Mankato, Minn.: Creative Editions, 2023.

## Websites

**Migration Videos**
https://thekidshouldseethis.com/?s=migration

**Where Do Birds Go in Winter?**
https://www.pbslearningmedia.org/resource/where-do-birds-go-its-okay-to-be-smart

**Wildlife Journal Junior | Arctic Tern**
https://nhpbs.org/wild/arctictern.asp

Every effort has been made to ensure that these websites are appropriate for children. However, because of the nature of the Internet, it is impossible to guarantee that these sites will remain active indefinitely or that their contents will not be altered.